Amazon Echo

Official 2017 Amazon Echo and Amazon Dot User Guide

Peter Gelson

Peter Gelson

© Copyright 2016 - All rights reserved.

the consent of the author or copyright owner. Legal action will be pursued if this is breached.

Table of Contents

Peter Gelson

Introduction

In the past 50 years, technology has made incredible leaps and bounds. It has not only made technology like televisions and phones more efficient and easier to use, but it has also introduced many new types of technology. This includes laptops, tablets, and having the cloud system, among many other things.

One of the many types of inventions that this book will focus on will be hands-free and voice recognition technology. This is incorporated in two new Amazon products: The Amazon Echo and Amazon Echo Dot.

These are wireless and hands-free speakers that use voice recognition to activate and use the speakers. The voice recognition system for these two Amazon products is Alexa. Through Alexa, you can use these speakers to do much more than simply play your music.

Throughout this book, the physical design and function of the two Echo devices will be fully fleshed out for you so that you can better use these devices on your own. There will also be examples of ways they can be used, how Alexa works, what commands you use with these devices, and what other users have said about the products.

There are many different ways you can use your Echo device, so this book is going to serve as a guide for you to follow and show you the ins and outs of the device. You don't have to follow each tip or know every part of your Amazon product, but it will help make it easier for you to find a way to use your Echo device so it suits your needs. There are literally thousands of things you can do with them, after all! Plus, there are a few hidden tricks you may not have been aware of that you could do with Alexa.

I'll start off by explaining the functions and designs of each device, then help you learn how to use your device with different commands and explaining what it can connect to.

Chapter 1
Amazon Echo Features and Specifications

The Amazon Echo has many amazing features for you to make use of and it has a pretty unique design. This chapter is going to break down the various physical features and designs of the Amazon Echo so that you can understand how your device works. It will also help you to compare the Amazon Echo with the Amazon Echo Dot.

The first eye-catching feature of the Echo device is definitely how it works in every direction. The speaker is rounded so that it sends out sound 360 degrees around it, which allows you to hear it no matter what part of the room you are in. A lot more speakers are now implementing 360-degree sound, but it is still a relatively new idea.

Another feature that goes hand-in-hand with this rounded speaker that can fill a room with sound is the use of seven microphones. This means that no matter where you are in the room, your device can pick up your voice, even when you are playing your music! In order to catch the attention of Alexa so you can begin using your Echo – to change the station, to use one of Echo's other skills – you have to begin by saying "Alexa." This will wake the device up so it can begin responding to what you are saying.

As I've already mentioned, these speakers can do so much more than just play music. When using Amazon Echo, you can activate various skills. I won't go into too much detail on them now, but these skills can range from ordering a pizza to calling up for an Uber ride.

Physically, the Amazon Echo is a pretty small speaker. It is about 9 inches (235 mm) tall and weighs around 37 ounces (1064 grams). This is very convenient for when you want to store the device, have a small speaker that doesn't take up too much space, or if you want to take the speaker with you on a trip.

There is a blue light ring on top of the speaker that will display the status of the Amazon Echo. This list tells what the light ring means in different situations:

- Solid blue ring with spinning lighter blue lights – Echo is starting up

- Lights are off – the Echo is on and waiting for you to give it a request

- Solid blue ring with cyan light facing speaker – Alexa is processing your request

- Orange spinning lights – Echo is connecting to Wi-Fi

- Solid red – the microphones are turned off (microphone button is on top of Echo)

- White light – you're adjusting the volume level

- Continuous spinning violet light – there was a problem connecting to the Wi-Fi

There is a lot that can be talked about when it comes to the Alexa Voice Service. One again, this will be better explained later in the book. But the main aspect of Alexa hat that I think is important is that it connects to the Cloud. This means that it will constantly adapt to your voice and speech patterns so that it can more efficiently complete the tasks you ask it to do. There will also be an explanation on how you can better secure your Echo and ensure that no one accidently orders something online. (You'll find out what I mean later on if you don't already.)

The Amazon Echo also has a feature that allows you to connect with other apps and Echo devices through Alexa. In short, if you have two Amazon Echo devices being used in your house and you wake it up by saying "Alexa," Alexa will respond through the device nearest to you.

There are also "accessories," as Amazon calls them, where you have the option to add them to your order so they can be used alongside your Amazon Echo! There's the Alexa Voice Remote, WeMo Mini, the Phillips Hue Starter Kit, which allows you to link all of your lighting, Insteon Connected Home Starter Kit, Warranty and Accident Protection, and the Mission Portable Battery Base.

The Alexa Voice Remote will be mentioned later in the guide, but it is essentially a remote with a microphone so you can use your Echo from a different room or if the room the Echo is in is too noisy. The WeMo Mini is a smart plug that also works through Alexa, allowing you to control whatever you have plugged into this smart plug (which must be plugged into a wall outlet to work, of course). This means you can use it to turn your devices off and on and use it like an automatic timer for your lights if you are going on vacation somewhere or won't be home for a while.

The Phillips Hue Starter Kit is essentially just smart lights. It allows you to place lightbulbs into whatever light fixture you want and to control them through Alexa. There is a bridge that comes in this starter kit that will allow you to use Alexa on the lights that are connected to it wirelessly. Insteon Connect Home Start Kit is more or less the same thing. The difference is that you plug something into the Insteon module – like your lamps or even a TV – and control it with, you guessed it, Alexa. You can dim your lights, turn them off and on entirely, connect to your thermostat, and control ceiling fans, among other things.

The Warranty and Accident Protection is fairly self-explanatory. It is essentially a coverage plan for your Echo device so that, if something damages it due to causes out of your control, you can get it replaced or fixed. The Mission Portable Battery Base is also fairly self-explanatory. This goes on the bottom of your Amazon Echo in order to charge it when you are not near an outlet, want to play music when there is a power outage, or simply forgot to take the charger with you. If you plan on having your Echo for a while or traveling around to different areas with it, these two are definitely good additional items to get.

Peter Gelson

Chapter 2:
Amazon Echo Dot Features and Specifications

The Amazon Echo Dot is very similar to the Amazon Echo device. This is because the Echo Dot performs many of the same functions as the Amazon Echo. There are two generations of Echo Dot, but they both perform many of the same functions because they both use Alexa. I'll note differences between the two devices when I feel it is important to note them.

The Amazon Echo Dot and Amazon Echo are both speakers and have the same voice recognition system. However, the Echo Dot is much more compact and has a few different physical features than the Echo.

The Echo Dot is a very small device when compared with the Echo. The Amazon Echo Dot stands at about 1.3 inches (32 mm) tall and weighs around 5.7 ounces (163 grams). This is a much lighter speaker for you to move around and takes up even less space than the Echo - at least in terms of height.

It's also different from the Echo because the sound comes from the top of the speaker instead of coming from all around the speaker. This is shown by Amazon also selling different cases for the second-generation Amazon Echo Dot.

You can also input sound directly into your Echo if you wish. You can hook it up to a stereo or, if you have the right connection for your phone, you can hook it up to your phone. It's just another variation on how you can play your music. It can also be used so you are using larger speakers to play the sound and only have the Echo Dot to use Alexa. And it is very helpful if you don't want your Echo Dot to be connected to the Cloud.

The Amazon Echo Dot has the same flashing light ring around the top. The flashing lights will mean the same thing the flashing lights on the Echo does, but I will include the information here again for your reading convenience.

- Solid blue ring with lighter blue lights that spin – Echo is starting up

- If the lights are off – the Echo is on and waiting for you to give it a request

- Solid blue ring with cyan light facing speaker – Alexa is processing your request

- Orange spinning lights – Echo is connecting to Wi-Fi

- Solid red – the microphones are turned off so Alexa or the Dot won't hear you (microphone button is on top of Echo)

- White light – you're adjusting the volume level. This can be done manually or via voice

- Continuous spinning violet light – Your Alexa or Dot isn't able to connect to the Wi-Fi

Also, as I did with the Amazon Echo, I'm including additional "accessories" you can get for your Echo Dot, two of which may come in handy if you intend on using it with other speakers. The accessories you can get include a second-generation only Echo Dot case, a 3.5 mm stereo audio cable, a 3.5 mm RCA adapter cable, an Alexa Voice Remote, and, of course, a Warranty and Accident Protection plan.

The case that you can get for your Echo Dot essentially cover the entirety of the device – minus the top – in whatever fabric and color you chose. Currently, you can get cases in charcoal, indigo, or sandstone colored fabric; or you can get it in merlot, midnight, or saddle tan colored leather.

The two types of cables mentioned are only useful if you want to hook up your Echo Dot to another set of speakers. You will have to check your speakers to see which one it requires or if either would even fit them. These cables are 4 feet long (1.2 meters long), so you'll have plenty of space for you to make use of both your Echo Dot and your speakers.

And, just as with the Echo, the same information about the Alexa Voice Remote and the Warranty apply. I won't repeat it here because that would be too repetitive.

Now that I have finished talking about the specific features of both the Amazon Echo and Amazon Echo Dot devices, we are going to move into how you set up both devices. We will also take a look at how to set up other devices they can sync or be paired with.

Chapter 3
How to Set Up Your Amazon Echo and Pair It with Other Devices

While the information in this next section will be likely be included in your quick start guide – which comes in the box with your Echo – I am going to go through each step on how to set up your Amazon Echo here. I'll also include some information about what you should do if you are having issues in any part of the process of setting it up or connecting to other devices, or if your Echo begins flashing different colors that mean it is having an error.

Step One: The very first thing you need to do before anything else is download the Alexa app. Don't worry about additional costs as this app is free. You can use it on phones and tablets (as long as they are at least Fire OS 3.0, Android 4.4, or iOS 8.0), as well as on a computer. Going to http://alexa.amazon.com with your browser (Safari, Chrome, Firefox, Microsoft Edge or Internet Explorer 10 or higher) will allow you to download and use the app.

Step Two: After you have created an account and signed in on the Alexa app, your next step is to turn on your Echo. Find a place near a power outlet where you want to put your Echo. Take your power adapter (that should be included in the box with your Amazon Echo) and plug it into your Echo, then into the outlet. The light ring will start as blue. When the light ring turns orange, your Echo will greet you.

Step Three: You have to connect your Echo to a Wi-Fi network. You start this process by opening up the Alexa app, selecting the left navigation panel, and choosing "Settings." From this menu, you will select "Set up a new device" if you are adding a new device to your account. If you are updating the Wi-Fi connection for this device, select "Update Wi-Fi." You'll then hold down the Action button on your Echo for five seconds, which will turn the light ring orange and connect your device to your Echo; a list of available networks will appear in the app.

You may be asked by the Alexa app to manually connect your mobile device to your Echo. Then you select your Wi-Fi, enter the password, and now you can use your device!

The default wake word to use Alexa and your Echo is "Alexa." This can be changed n the Alexa app. You can change it to another name, a set of numbers, or even a random word that you will remember.

This will be explained and added onto throughout the book, but there are many things you can ask of Alexa through your Echo. You can check your calendar, control your music, control your smart home devices, hear the news, listen to audiobooks, set up/cancel timers, and many more things! You connect through the Alexa app and by enabling particular skills. As of the time I wrote this book, there are over 10,000 skills available for you to enable!

There are a few other things you can use alongside your Amazon Echo, such as your mobile device or Bluetooth-enabled devices.

Echo and Your Mobile Device

The first device I will talk about is something many of you already have: a mobile device. Because your Echo has Bluetooth connectivity, you can stream audio services, like iTunes, from your mobile device and through your Echo. It won't read off any text messages or notifications you receive, and it can't take calls, so this is only useful if you want to use a service outside of Amazon to stream your music. However, the bonus to connecting your mobile device and your Echo is you can also play any music you have downloaded onto your mobile device or tablet.

Here is how you connect the two devices:

> Step One: Make sure your mobile device is set to Bluetooth pairing mode and that it is within the range of your Amazon Echo. Once that is done, say "pair." Alexa will tell you that Echo is ready to pair devices. If you want to exit this process at any time from here, just say "cancel."

> Step Two: Open the Bluetooth Settings menu on your device and select your Amazon Echo. Alexa will let you know if the two devices are successfully paired. Once they are, you can stream audio straight from your mobile device through your Echo.

> Step Three: When you want to disconnect the two devices, say "disconnect." After they have been paired once, you can immediately connect the two any time after that by simply enabling Bluetooth on your mobile device and saying "connect."

And there you have it! Your mobile device can now play music through apps like Google Play Music on your Amazon Echo.

Echo and the Alexa Voice Remote

This is an additional item that you have to buy separately from your Amazon Echo. The Alexa Voice Remote is used to speak with Amazon Echo devices and can only be paired to a single device at a time. If you want to use the voice remote and you have several Echo products, you need a separate remote for each.

Before I explain how to pair these two devices together, you might be wondering what the point of the remote is? The remote makes it possible to use the microphone included within the remote to continue using your Echo if it is too noisy for you to be heard or you are too far away. It also functions as a regular remote with the standard volume controls, pause/play buttons, and previous/next buttons. If you want to be able to use your Amazon Echo at almost any distance within your home, maybe this remote will help you out.

Without further ado, here is how you connect your Alexa Voice Remote with your Amazon Echo:

Step One: You need to insert two AAA batteries into the remote – which are included in the box with the remote, no worries – before you can move onto anything else.

Step Two: Open the Alexa app, select the left navigation menu, and select "Settings."

Step Three: Select the Amazon device you are pairing the remote with (in this case, your Amazon Echo) and then select "Pair Remote." If you already have a remote paired with your device and you are trying to pair a new one, you must select "Forget Remote" first.

Step Four: Press and hold down the play/pause button on the remote for five seconds. Your remote should be paired with your Echo within a minute. When the Echo and the remote have been paired, Alexa will let you know by saying "Your remote has been paired."

And now you have your voice remote paired with your Amazon Echo! A cool advantage of this is that you don't have to use the wake word and it works even if your Echo is muted. You can change this at any time through your Alexa app by selecting "Forget Remote."

Echo and Bluetooth Speakers

Having these two paired up may seem like a strange combination. Why would you need other speakers to stream music from a speaker? But this is helpful for those who have their Amazon Echo on anyway and play music through something like Spotify. Having your Echo connected with Bluetooth speakers means that you can take the music you want to play from your Amazon Echo and have it actually play through your Bluetooth speakers.

So here is how you connect the two devices:

> Step One: Set your Echo and Bluetooth speaker about three feet apart. This will make it easier for Alexa to hear the wake word. Make sure your Bluetooth speakers can connect to other devices, like a smartphone. Once you are sure it can connect, turn the Bluetooth speaker on and put the volume up. **Keep in mind that you can only connect a single Bluetooth device at a time**, so other devices must be disconnected.

Step Two: Turn on pairing mode for your Bluetooth speaker. Each device will have a different way of doing this, so check the user guide that came with your speakers to figure out how to do this. Once this is done, open up the Alexa app and select "Settings." Select your device, followed by "Bluetooth," then "Pair a New Device." When they are connected, the speaker will appear in the list of available devices within your Alexa app.

Step Three: Select you Bluetooth speaker so the Amazon Echo can connect to the speaker. Alexa will let you know if the connection was successful. In the app, select "Continue" and your speakers will officially be connected to one another. If you want to play audio from another device, say "Disconnect."

So now you have all this information on how to set up and pair your Echo, which is great! But what if you run into trouble? Here is one basic solution to help you out along the way.

Reset Your Amazon Echo

If you are having issues with your Amazon Echo or first-generation Echo Dot, first try restarting your device to see if that will fix it. (Turning it off and on again to make it work is handy in almost any situation!) To restart your Echo, you'll first need to unplug the power adapter from either the Echo itself or from the wall outlet and then you'll need to plug it back in.

If that doesn't work, follow these steps to reset your Echo device:

> Step One: Use a paper clip or another object with a small point to press and hold the Reset button on your Echo for five seconds. After this, the light ring on the Echo device will turn orange and then blue.
>
> Step Two: Now you have to wait for the light ring to turn itself off and on again. The light ring will turn orange as your Echo enters setup mode.
>
> Step Three: This step involves going back through the setup process (connecting it to Wi-Fi and registering the Echo to your account). It's frustrating to go back through, but this time around you'll know what to do at least!

There's only one within this section, but check out the chapter on Alexa (at the end, just like this chapter) to see if any of that information helps. This will mean that, if you are having an issue using Alexa or with streaming audio, Alexa is actually the part that is having problems.

In any case, that's all the information I have for now on the Echo. So let's move onto the Echo Dot! (Got An Amazon Echo For Christmas? Here's How to Set It Up)

Chapter 4
How to Set Up Your Amazon Echo Dot and Pair It with Other Devices

Just as with the Amazon Echo, I am going to discuss how you set up your Echo Dot. Many of the steps are the same. This means that a lot of what I am going to write in the next few paragraphs is mainly repetition if you read about how to set up your Amazon Echo.

Now that you have your warning, here is how you set up your Echo Dot:

> Step One: Download the Alexa app onto your phone or tablet or access it through your browser and computer. You will need to create or sign in to your Alexa account.

> Step Two: Turn on your Echo Dot. It is recommended that you place it about 8 inches away from any wall or window, but it is up to you where you choose to place it. Plug the power cord into your Echo Dot and then

into the wall outlet. The light ring will begin as blue and turn orange when your device is ready to be used.

Step Three: You have to connect your Echo Dot to your Wi-Fi network. Open up the Alexa app, select the left navigation menu, and then choose "Settings." Here you will then click on "Set up a new device," if the device is new to your account. If it is not a new device, select "Update Wi-Fi." You will hold down the action button on your Echo Dot and wait five seconds. The light ring will change from orange to blue when it is ready and Alexa will alert you that you can now use your Echo Dot.

And there you go! Now it is all set up. When you want to wake it up and use it, use the wake word. The default wake word is "Alexa," but you can change this within the Alexa app.

There are a few minor differences that I would like to note now in order to avoid confusion between the two generations of Echo Dots.

For a physical volume control button on the device, they have two slightly different controls. The second-generation

Echo Dot has an up and down volume control button. This is the standard volume control button for many devices. By comparison, the first-generation Echo Dot has a volume ring. You control the volume on the first-generation Echo Dot by turning the ring clockwise to turn the volume up and turning it counter-clockwise to turn the volume down. The volume ring on the first-generation Echo Dot functions just like the volume ring on the Echo.

Also, the first-generation Echo Dot has an LED power light and a specific reset button on it. The second-generation Echo Dot needs a button combination in order to reset it and doesn't come with an LED power light.

As you can see, these are very small and minor differences between the two devices. They don't impact much aside from how you physically deal with the device and, even then, it doesn't have an incredibly large impact. But now it is time to move on!

Echo Dot and an External Speaker

You may remember that I mentioned that the Echo Dot can be connected to another speaker. You can do that either through Bluetooth or with an audio cable compatible with your Echo Dot and your speakers. The purpose of connecting the two would together would be so that you could still use hands-free technology to change and play your music while having it on better or more speakers that you already have set up.

I will explain how you connect your Echo Dot through both Bluetooth and hardwire options.

Let's begin with the audio cable connection to an external speaker:

> Step One: Make sure that you set your Echo Dot and speakers at least three feet away from one another. This is important to make sure that Alexa can still hear you and that you request won't be drowned out by the music playing.

> Step Two: Turn your external speaker on and plug an end of the audio cable into the Echo Dot and your speakers.

And there you go! Now any music you play from your Echo Dot should be projected through the external speaker. Please note that if your speaker doesn't have a 3.5-mm audio port, you can use audio cable adapters until it is the right size to fit into your speakers.

Next is how to connect your Bluetooth speakers with your Amazon Echo Dot.

Step One: Just as you did with your external speakers that required an audio cable, make sure you set the two devices at least three feet away from one another. You should also make sure that your speakers can connect with other Bluetooth devices and make sure you disconnect any other Bluetooth devices from your Echo as it can only connect to one Bluetooth device at a given moment.

Step Two: Turn on pairing mode for your Bluetooth speaker (check your user guide to see how you do this) and go to your Alexa app. Once there, select "Settings."

Step Three: Select the Bluetooth speakers and choose "Bluetooth" followed by "Pair a New Device" within your Alexa app. Once your Echo Dot can find the speakers, it will appear in your list of devices within the Alexa app. Select your Bluetooth speakers one more time to completely connect your Echo Dot with the speakers. Alexa will tell you if they pair up properly. Anytime you want to disconnect your Echo

Dot and your Bluetooth speakers from one another, just say "Disconnect."

There you have it! Now you know how to connect external speakers with your small Echo Dot.

Now it's time for some solutions to problems you may run into with your Echo Dot. Remember that some problems may actually be a result of an issue with Alexa, so check over there if you don't see and answer to your questions here.

What to do if your Echo Dot doesn't turn on or respond at all
This can occur primarily with your Echo Dot because you can use other adapters with it and because you can connect it up with other speakers. Using the wrong cable, having the speakers too close to one another, among other things, can cause problems. These aren't possible with the Amazon Echo.

So here is what you should check if your Echo Dot won't turn on or respond:

• Make sure you use the included 9W power adapter so your Echo Dot gets enough power supplied to it. For example, cell phone chargers will not provide enough power for your Echo Dot to work.

- Press the **Action** button and see if Echo Dot responds to your requests.

- When you connect your Echo Dot to a speaker, make sure your Echo Dot is at least three feet away from the speaker so it can hear the wake word as well as the requests you make of it.

- Make sure the Echo Dot is at least eight inches away from other objects.

- Try to avoid having background noise when you make a request.

- Speak naturally and clearly when making a request.

Reset your second-generation Echo Dot

If you read through the chapter on the Amazon Echo, you know that the first-generation Echo Dot is reset the same way as an Amazon Echo. If you have a first-generation Echo dot, go look back at that section instead. Here are the steps to reset your second-generation Echo Dot:

Step One: Press the "Microphone off" and "Volume down" buttons at the same time and hold them down until the light ring turns orange. This should take about 20 seconds. After this, the light ring will turn blue.

Step Two: Wait for the light ring to turn itself off and back on again. Once it is back on, it will be orange once again as it enters setup mode.

Step Three: Open the Alexa app to connect your Echo Dot to a Wi-Fi network and re-registering it to your Amazon account.

What do I do if my Echo Dot isn't connecting to another speaker?

Echo devices can connect to many different speakers both over Bluetooth and by using an audio cable. Keep in mind that speakers that require a PIN to pair the devices will not be supported by your Echo Dot. Now that I have made note of this, let me take you through the steps you should take if it won't connect to your speakers.

If your Echo Dot isn't connecting to a Bluetooth speaker, here is what you should check and things you should try doing:

- Try to connect another device, like your smartphone, to the speaker.

- Make sure there are no other devices paired with this speaker.

- If it is a portable speaker, check the batteries and maybe replace them.

- Make sure both your Bluetooth speakers and Echo Dot are away from object that can cause interference, like a microwave oven.

- In the Alexa app, forget the Bluetooth speaker by going to your "Settings," selecting your Echo Dot, and then selecting "Bluetooth." Here, you will be able to select your Bluetooth speaker followed by "Forget Device." This will unpair the two devices, so you can try re-pairing them and see if that helps.

If your Echo Dot isn't connecting to a speaker while you are using an audio cable, here is what you should check and things you should try doing:

- Make sure the sound ports are clear and that nothing could be blocking hem.

- Make sure that you are using the proper size cable (3.5mm). If you aren't, the sound won't go through.

- If it is an older stereo system, try using an adapter.

If none of these solutions solves your problems, check out the section about Bluetooth connectivity problems with Alexa.

We've covered a lot of the physical features and basic set up of both the Amazon Echo and Amazon Echo Dot. It's time to move onto how Alexa functions. (Amazon Help)

Chapter 5
The Alexa App: What It Is and How It Functions

There's a lot of information about the Amazon Alexa app that I will be explaining in this chapter. However, as of right now, you can only download the Alexa app from US app stores. Keep that in mind as you try to set up your Amazon Echo.

The Alexa app is available on certain mobile devices. This includes systems like the Fire OS 3.0, Android 4.4, or iOS 8.0 and higher OS systems. If you are using a computer through Wi-Fi, you can also download it through your web browser. I briefly mentioned this information when I explained how to set up your Echo and Echo Dot, so that's all I will mention here.

Now I'm going to talk about the basics of the Alexa app itself. There are two main things within your Alexa app that will help you navigate it: The home screen and the left navigation panel.

Peter Gelson

The home screen will show your activity with Alexa. Here, you can also scroll through the "Cards" to see descriptions of your recent requests.

Then there is the left navigation panel. Select the menu icon with Alexa app to open it. This panel will allow you to access many different Alexa features and settings. If you are viewing the Alexa app on your computer, this is automatically open for you. This is where you can access and manage all of the various kinds of things that you create with Alexa through your Amazon Echo and Echo Dot.

Within this left navigation menu, you can select: Home, Now Playing, Music & Books, Shopping & To-Do Lists, Timers & Alarms, Skills, Smart Home, Things to Try, Settings, Help & Feedback, as well as Not ...? Sign out.

This is where you will access and manage many of the things you create using Alexa with your Echo device. Many of these are fairly self-explanatory or I will get into them in more depth in the next chapter, so I won't discuss most of them here. "Things to Try" is a section that has a list of phrases that you can try with Alexa. "Settings" is probably the menu you will use the most, as you can adjust your Echo's settings, among other things, within this menu.

Within the "Settings" tab, you will find settings for your Echo devices by selecting your device. These settings include: Update Wi-Fi, Bluetooth, Pair device remote, Sounds & Notifications, Device name, Device location, Wake word (where you can change Alexa's wake word), Temperature Units, Distance Units, Device is registered to: , Device software version, Serial number, and MAC Address.

I have already talked about pairing a remote with your Echo, so I won't go into that once more. I'll talk about "Sounds & Notifications" and "Wake word" within this chapter and the next chapter, so you can get the most out of your Echo device.

"Temperature Units" and "Distance Units" are helpful so Alexa can report information to you in the units that you understand. For example, turning on these two options will result in Alexa measuring in Celsius and meter-units. "Device location" is also helpful if you want easy-access information about the weather, the current time, places nearby you, or things of that nature. You simply input your address and select save.

As I've already mentioned, you can change the wake word for your Echo and Echo Dot. Right now, there are only a select few words that you can use as a wake word. Current options you have for wake words include "Alexa," "Amazon," "Echo," and "Computer." This allows you different ways to wake up your devices in a way that won't cause any confusion. For example, if someone in your household or a friend has the name Alexa, your Echo device may try to wake up and process a request that they can't understand because it wasn't actually directed towards them. You will have to remember this wake word each time you go to activate your device or to make use of a skill, so keep that in mind.

There are more interesting things you can do in "Settings" in addition to looking up information about your Echo device. In your account settings, which you get to by simply selecting the "Settings" menu and not clicking on a device, you can select Voice Training, Music & Media, Flash Briefing, Sports Update, Traffic, Calendar, Lists, Smart Home, Voice Purchasing, Household Profiles, and About the Amazon Alexa app.

I'll be going into detail at least slightly on how you can access your calendar, manage your lists, purchase items, and listen to music in the next chapter. For now, I'll be explaining the Voice Training, Flash Briefing, and Smart Home options.

Voice Training

Voice training will allow you to improve the speech recognition of your Alexa device. Within voice training sessions, you will see 25 different phrases within the app to repeat to your Alexa device. Your device will process each phrase you say. The best results during these voice training sessions will occur when you speak normally to your Alexa device and sit or stand where you would normally be while using your Echo or Echo Dot.

To begin a voice training session, here are the steps you must take:

Step One: Open the Alexa app, select your left navigation panel, and then select "Settings." Once you are in your settings, select "Voice training."

Step Two: You can now click on "Start session." You will speak the phrase that appears in your Alexa app and then select "Next" when you are finished. If you need to repeat the phrase, select "Pause" and then "Repeat phrase." Once you've reached the end of your session, hit "Complete." If you need to end your session early for whatever reason, just select "Pause" once again and click on "End session."

So why would voice training be important to you as an Alexa app user? It's important because Alexa connects with the cloud in order to constantly improve its voice recognition system. With enough voice training or enough usage, Alexa gains a better understanding of your voice patterns so that it will more easily understand what you are requesting it to do.

If you want to continue improving Alexa and its AI system, this is a good skill to use. It may also help you remember to enunciate much better than you would in your daily life to request something from Alexa. For example, you may often have your voice drop off near the end of a sentence. This is something humans can register but it is still difficult for an AI to pick up on these same speech nuances. The bonus to Alexa in this case is that Alexa is always learning.

Flash Briefing

I just wanted to quickly add this section in here, although it may be better within the next chapter.

Flash briefing is just a type of command you can give your Alexa device. This will quickly update you on shows, news headlines, and weather you choose to select within the appropriate tab in the "Settings" menu. You can connect with many news and weather sites within flash briefing.

Essentially, asking "Alexa, what's new?" or "Alexa, what's my flash briefing?" are the ways you can get a very quick overview of what is happening in the news and weather and on shows.

Smart Home

I will elaborate further on smart home devices in the next chapter. Here, I will just explain what they are very briefly. There are particular smart home devices that can pair up with Alexa. Many of these, although not all, are Bluetooth-style smart devices; however some work with your wi-fi connection. They typically plug into a wall or have a bridge that connects multiple types of the same thing (like light bulbs). The wall outlet smart devices allow you to plug another device into it to control it remotely. This is especially common with lights, where you can dim and turn them off and on as you want.

Some of you may have smart home devices automatically built into your home. These may still be compatible with Alexa! You will have to check them and see if you can use them with your Alexa app.

Alexa Skills

Another important thing about the Alexa app is the Alexa skills. I have already gone into some detail about them while I was talking about the Echo and Echo dot. An Alexa skill is sort of similar to an application on your phone. It's the equivalent of having an Internet browser, your favorite music station, and occasionally even games on your phone, but you can just use them hands-free with your Echo and Echo Dot. As I have already noted, there are over 10,000 skills currently available for Alexa to use. You can search for skills within the Alexa app or from the Alexa skills store on the Amazon website.

It's very simple to use these skills and enable them, but it may be difficult for you to figure it out on your own. This is why I will explain it on a case-by-case basis, albeit with very few examples, in the next chapter.

To enable a skill for Alexa through the app, you will need to select your left navigation panel, go to the "Skills" tab, and use the search bar or categories button to find the skill you want to use. Once you have found the correct skill, you simply select "Enable Skill." If you are using the Amazon website, you do the same thing, although the panel should already be open at the start. Once a skill is enabled, you can ask Alexa to open and use that particular skill whenever you want to use it. Some skills will list commands that are recommended for you to use whenever you want to access it.

I also highly recommend checking out the reviews of each skill. There are many third-party skills that Alexa can enable. This is great because it provides more variety, but some may not work well or respond well to the commands listed for them. It's a matter or trial-and-error in some cases. But it is worth looking into to see if that particular skill is worth enabling.

There's one more feature of Alexa that I'm going to go into right now and it is about the "conversations," if you want to call them that, that you have with Alexa.

Viewing your Dialogue History

New conversations with Alexa are recorded whenever you have awakened it. One thing Amazon mentions on their help page for this exact topic was that deleting these recordings may lead to lessening your experience with using the Echo or Echo Dot. The Alexa app revolves around using old information in order to grow, so deleting conversation can impact how well it can understand what you are telling it to do. In any case, here's how to listen to your dialogue as well as how to delete your interactions with Alexa through your Amazon Echo and Echo Dot:

- To listen to past recordings: Select "Settings," followed by "History," in the Alexa app. You can then select a recording from the list and press the play button to listen to it.

- To delete a single recording: Select a recording, follow by "Delete voice recordings." This will remove the audio from the Cloud and from any of the Home Screen cards within the Alexa app. **Side note:** If you want to delete a card from your Home Screen, go to the Home Screen and select "Remove card."

- To delete every recording you have ever had with Alexa: Go to "Manage your Content and Devices," or www.amazon.com/mycd, and select the tab marked "Your Devices." Here, you select your Echo or Echo Dot followed by selecting "Manage voice recordings" and "Delete."

And there you have it! Hopefully, you can now feel a little more secure while using your Echo or Echo Dot. If you want to feel safer and make sure it won't pick up your voice and conversations when you don't want it to, you can always turn off the microphone by using the "microphone off/on" button that is on the top of both Echo devices. If you don't plan on being near your Echo device or you're fine with manually changing your music through the Alexa app, these would be good times to turn the microphone off. Now it's time for the last part of this chapter...

Household Profiles

I haven't mentioned this in the book so far, but you can actually add another person to your Amazon Household so you can both use and listen to one another's content, like music or calendars. If they are to make a purchase, however, make sure that they are either using their own Amazon account to complete the purchase or that they will pay you back. By adding them to your Household, you are allowing them access to make purchases with your card.

Now that you have a little background and a small warning, here is how you manage Household profiles:

- To add someone to your Household: After opening the Alexa app, go to the left navigation panel, select "Settings," go to "Account," and finally press "Household

Profile." Alexa will give you instructions from here as to how you allow a second person to join your Household. Make sure they are present when you are inviting them because they will need to enter their information now.

- To remove a person from your Household: In the Alexa app, go to the left navigation menu, select "Settings," then press "Account" and "Household Profiles." You can either select "Remove" to remove another person or select "Leave" to remove yourself. Selecting "Remove" from Household will confirm that they are no longer in your Household. **Please note:** When you remove a user from a Household, you can't add either yourself or the other person to another Household for 180 days. If someone is accidentally removed, contact Customer Support.

- To switch between profiles: After waking up your Echo or Echo Dot, say "Switch accounts." If you want to check what account it is currently one, ask "Which account is this?" If you are browsing a content library (like a music library) within the Alexa app, use the drop-down menu to switch between your libraries.

What if Alexa won't work?

This is targeted primarily toward the Alexa app not functioning on whatever device you have chosen to use it on. Your issue could be that your device isn't compatible, so make sure to double-check that first and foremost.

If it is meant to be compatible, here is what you should do for each type of device:

- For your iPhone, iPad, or iPod touch, turn the device off and on again. You should also force close the app. To do this, press the Home button twice until you see a list of your recently used apps. Close the app by swiping up. You can also uninstall and reinstall the app to sort of restart the app itself.

- For an Android device, turn it off and on again. Once again, you should also force close the app. To do this, go to your "Setting" menu and select either "Apps" or "Applications." For some devices, you might have to select "Manage applications" next. Once this is done, find the Alexa app on your list of installed apps and press it. Here you can select "Clear Data," followed by "Force Stop." And you can also uninstall and reinstall the app.

- For a Fire OS device, turn it off and on again. If this doesn't work, force close the app by going to your Home screen and swiping down from the top. This opens "Quick Settings," where you can then select "Settings" or "More," followed by "Apps & Games," or "Applications," followed by "Manage All Applications." All of these options should let you find the Alexa app from the list of apps you have installed. Once you have found it, select "Clear Data" and then "Force Stop." As with the other devices, you can also uninstall and re-install the Alexa app to let it restart if all else fails.

- If you are on your browser, reload the webpage. If it's still not functioning, clear the cache and cookies from your browser. Keep in mind this will remove any remembered website settings you may have, like your usernames and passwords. Also try closing and re-opening your browser again.

We've covered everything that could be covered about Alexa that would fit within this chapter. But this wasn't *really* the information many of you reading were looking for. No worries! This upcoming chapter is all about Alexa's skills and commands, where you will hopefully find the information you want.

Peter Gelson

Chapter 6
Alexa Skills/Commands

Alexa is such an interesting type of system to use because you can do so many different things with it! With all of the useful ways you can make use of it, along with the fun additional features, you have a seemingly infinite number of ways for you to make use of Alexa! I'll list only a few here — going through each and every skill you can enable or doing an example of each would take up a whole book on its own!

I'll make a list so you can get a small understanding of what skills — which, I should mention, are basically like different apps — you can use with Alexa (there are currently 10,000 available!) on your Echo or Echo Dot devices. You can open skills you enable through different commands, which is what the list will be made up of.

- "Alexa, open Starbucks."

- "Alexa, turn on the lights."

- "Alexa, open Rain Sounds."

- "Alexa, set a timer."

- "Alexa, open Top Ten Countdown."

- "Alexa, open weird facts."

- "Alexa, ask Fitbit how I'm doing today."

- "Alexa, what's in the news?"

- "Alexa, what's my Flash Briefing?"

These are just a few of the commands you can use with your Alexa app. There are so many different ones available to you, depending on what skills you enable. You can always check Amazon to see what new skills are available for you to use and what commands should be used for each skill!

Before we dive into these various skills and commands, let me briefly explain how having multiple Alexa devices works. When you have several Alexa-run devices (multiple Echo or Echo Dots, for instance), Alexa will respond through the device you are closest to with what Amazon (humorously) calls ESP – or Echo Spatial Perception. Please keep in mind that you can't interconnect your Echo or Alexa devices together to play the same audio or something like that. If you want to use a particular device, you can give your device a unique name.

You will, however, get to share some of the same content and setting between your devices, such as music, Household profiles, smart home devices, and shopping lists. These should be the same when you look at them in the Alexa app across all of your devices. You don't get to share the alarms, times, sounds, or Bluetooth connections between devices, however. This will be where you would have to use each device separately.

Also keep in mind that Alexa will respond any time it hears "Alexa" or whatever you have changed the wake word to. "Alexa" and "Computer" are likely the worst wake words as they are used so often in everyday life. This means that having a TV show or ad say "Alexa" or "Computer" will wake your device up and begin recording. Keep this in mind when you have an Alexa-controlled device in your home. And remember that there is a "Microphone off" button on your device.

Now that we've covered that, it's time to begin going into more common commands you may find useful while you are using Alexa! Starting with...

Playing Music with Alexa

This section will require you to enable or connect your music account to Alexa. There are many different music services you can use on your Echo or Echo Dot, such as Amazon Music, Prime Music, Spotify Premium, Pandora, TuneIn, iHeartRadio, and Audible. You can ask Alexa to directly stream music or other media from these services.

There are some streaming services that will require you to link your account with Alexa, including Spotify Premium, iHeartRadio, and Pandora. To link these with your device, go to "Settings" within the Alexa app. Then select "Music & Media," tap whichever music streaming service you with to use, and finally hit "Link account to Alexa." You will then see a sign-in page appear in the app where you enter your information for the music service. And there you go! It's all linked up.

If you want to upload your own music from services like iTunes, use Amazon Music on your computer and upload your music to the collection called "My Music" on Amazon. There is a limit to how many songs you can upload to "My Music," but any songs your order from Amazon's Digital

Music Store won't count towards this limit. You can have up to 250 songs on this list.

However, if you are an Amazon Prime member, you will have a lot more music available to you. With Amazon Prime you can take advantage of Amazon Music's streaming service. Much like Pandora, Amazon Music allows you to stream music on your phone, tablet, TV, or Echo devices. You can stream genres, specific artists, or specific albums. Once your Amazon Prime account is linked to your Alexa, you can ask Alexa to play any of the stations you created or any of the stations that are already available to you on Amazon Music.

Another feature of Amazon Music is the playlists. When you use your phone, tablet, or computer to log into your Amazon Prime account, you can use Amazon Music to create endless playlists. Amazon Music also has an affordable subscription service that broadens your music availability to even the most recent released music albums.

In addition to online streaming services, you can connect to your Amazon Echo or Amazon Dot and stream music right from your device. This allows you to play any album or playlist you have saved in your phone or tablet without physically hooking up to your Amazon Echo or Amazon Dot. Just turn on the Bluetooth capability on your device and tell Alexa you want to connect to it. Alexa will walk you right through the process.

As sort of an aside, if you are searching for a song, genre, playlist, or artist that isn't available in "My Music," Alexa will begin to search the Amazon music catalogue or for a sample from the Digital Music Store.

Now I'm going to list a few basic commands you will be using when you are playing music on your Amazon Echo or Amazon Echo Dot with Alexa:

- To adjust the volume, say either "Volume up," "Volume down," or "Set volume to level..." ("Set volume to level 4," for instance).

- To hear the details of the song that is playing, you can ask, "What is this?" "Who is this by?" "What song is this?" or "When did this (song or album) come out?"

- To stop the song playing, simply say "Stop" or "Pause."

- To continue playing music, say "Play" or "Resume."

- To change to a different song, you can say "Next" or "Previous." This particular command is available in TuneIn. You can't go back a song in Pandora or iHeartRadio.

- To loop the music list, say "Loop." This command isn't supported in TuneIn, iHeartRadio, Pandora, Spotify Premium, or on Amazon Stations.

- To shuffle all of the songs together, say "Shuffle." To undo this feature, say "Stop Shuffle." This command isn't supported in TuneIn, iHeartRadio, Pandora, Spotify Premium, or on Amazon Stations.

- To repeat songs, say "Repeat." This command isn't supported in TuneIn, iHeartRadio, Pandora, Spotify Premium, or on Amazon Stations.

There is also an option to set a sleep timer to stop playing music after a particular amount of time. You set this up by saying, "Set a sleep timer for..." ("Set a sleep timer for 3 hours") and after that period of time, your music will turn off.

As you will have noticed, not all commands are available on each type of music app you may want to play music on.

However, this shouldn't be much different than how they are normally used anyway, so it shouldn't have too large of an impact. Just be aware that certain commands may work with one skill, but not with another.

Now that you know the basics of how to play your music, here are some more advanced commands that should work no matter what app or radio system you are using:

- To play a song, you can either say "Play some music" or "Play the song..." to request a specific song to be played. ("Play the song, 'Starboy.'")

- To play an album, you say "Play the album..." ("Play the album, 'Death of a Bachelor.'")

- To play music only by a particular artist, tell Alexa "Play songs by..." ("Play songs by Nikki Minaj.")

- To play songs based on the genre of music, ask Alexa to "Play some... music." ("Play some rock music.")

- To play a playlist you have on your account with whatever music service you use, say "Listen to ... playlist." ("Listen to My Music playlist.")

If you get can't remember the name of a song you want to play, you can still ask Alexa to play the song. You ask by saying, "Alexa, play the song that goes '...'" ("Alexa, play that song that goes, 'Young bull living like an old geezer...'"). This may not work all the time, particularly if there are multiple covers or editions of that same song. I'm also not sure that Alexa would be able to find an instrumental song, since you can only hum out how the song sounds.

There are other commands you can use for each individual service I have mentioned here, such as "Play some Prime Music," so you can use each service when you want to. There are so many commands and several services I would need to run through to cover them all, so I am not going to go into detail on them here. Instead, I'm going to briefly discuss another major use for your Alexa device.

Listening to Audiobooks and Podcasts
This may not be as popular as some of the other skills I will go into detail about, but I thought I would add it because it is a very helpful function. You can listen to audiobooks through Audible and Kindle Unlimited on your Echo devices. There are features from Audible that Alexa doesn't support, including reading newspapers or magazines,

creating bookmarks, or controlling the narration speed. Now that the introduction is out of the way, let's look at the commands you can use to listen to your audiobooks:

- To listen to an audiobook, say something along the lines of "Read..." or "Play the book ..." ("Read 'Paper Towns'" or "Play the book, 'Paper Towns'" are both valid, for instance.)

- To pause the audiobook, simply say "Pause."

- To continue the last audiobook, you listened to, say "Resume my book."

- To move forward or backward in the book by 30 seconds say "Go forward" or "Go back."

- To switch between chapters, say "Next chapter" or "Previous chapter."

- To go to a specific chapter within the book, say "Go to chapter ..." ("Go to chapter 4").

- To restart a chapter, say "Restart."

Just as with your music, you can set a sleep timer to stop reading a book after a certain period of time using the same command. This will help to make sure you don't walk away from your device and come back hours later with the book entirely finished without you having been there.

This can be a great idea if you are working on a new DIY project or following a recipe. You can have Alexa read it off to you so you don't have to dart constantly between whatever you are doing and the book.

I included podcasts in this section because it involves many of the same controls as reading an audiobook. You can use TuneIn to listen to various podcasts and just ask Alexa to play your favorite! Keep in mind that if you ask Alexa to do something else while you have it paused, you will lose your place in your book or podcast. Also, you automatically listen to the latest podcast within a series, so you would have to continuously say "Alexa, play the previous episode" if you wanted to go back further.

Now you know how to listen to your audiobooks on your Amazon Echo and Echo Dot devices! Let's move forward onto another good skill Alexa can use...

Wikipedia Surfing

That's right, just as many of us end up doing when we get either too bored or don't know where to begin searching for a topic, we can listen to Wikipedia through Alexa. What's fun about this is that you can ask Alexa to look up a topic and it will search for an answer. Or you can specifically ask to go on Wikipedia for a particular topic. ("Alexa, Wikipedia: 'Fantasy Football.'") Once you've listened to the beginning of a Wikipedia article, you can say "Alexa, tell me more" to hear more of the information within said article. Pretty interesting way of hands-free, and potentially very time-consuming, looking up information.

"Looking" at Your Calendar

Before anything else, let me say that you can link one of each type of calendar to Alexa through your Amazon account. Another person in your Household can do the same. The supported calendars you can use are Google Calendar and Microsoft Calendar. These will be synced up with your Echo, which anyone can access, so keep that in mind when you connect the two.

You connect your calendar to Alexa by going to the "Settings" menu, selecting "Calendar," and selecting the calendar you want to link to Alexa. Once you have done this, you click on "Link calendar account" and sign in with information you use for that calendar. Now you can access information about your calendar using your Echo Devices! As I am doing with all the skill I am including, here are some commands to use:

- To find out when your next event is, ask either "When is my next event?" or "What's on my calendar?"

- To find out about an event at a particular time or day, ask something like "What's on my calendar tomorrow at ...?" ("What's on my calendar tomorrow at noon?") or "What's on my calendar ...?" ("What's on my calendar Friday?")

- To add an event, you can simply say "Add an event to my calendar" and answers prompts from Alexa. You can also say "Add... to my calendar for..." ("Add 'Lunch with Tyler' to my calendar for noon tomorrow").

Now you can be on top of your schedule at all times! So long as the microphone and your Echo device are on, of course. Now that we have covered your calendar to keep your schedule intact, let's go on to how you can use and create timers and alarms through Alexa.

Timers and Alarms

This is a handy feature for you to make use of, especially if you are planning to play music while doing housework, while you're studying for a test, or while you're cooking. One thing to note is that your timer and alarms won't sync up between your Alexa devices; they will be made on each device separately.

You can also manage your timers through your Alexa app and set up a single timer or alarm up to 24 hours in advance. So you can't set a countdown timer for an event that's a week away (sadly).

Each Alexa device uses timers and alarms independent of other Alexa devices registered on your Amazon account. Use your voice to set up or cancel a timer or alarm, and you can then manage them in the Alexa app. You can set a single timer or alarm up to 24 hours ahead. However, you can set up to 100 alarms and timers if you want to!

Now that you have a little bit of information to begin, here is how you set up and use a timer with Alexa:

> Step One: To set up a countdown timer, you will say "Set a timer for..." and then say how long you want the

timer to be going for. This could be five minutes, 4 hours, 24 hours, whatever you want.

Step Two: Once it has been set, you can ask Alexa how much time is left on the timer or ask to stop, cancel, or resume a timer.

Step Three: If you have multiple timers set, you will need to use the Alexa app to manage them. To do this, you select "Timers & Alarms" from the menu, select the device with the timer on it, and click on the "Timers" tab. Here you can select "Edit" next to a timer you want to manage if you want to pause or cancel a timer.

And there you have it! Nice and easy for you to use. Now here is how you set up and use an alarm with Alexa:

Step One: You can set an alarm by saying, "Set an alarm for..." followed by the time of day you want it to go off at ("Set an alarm for 3 pm"). If you want to set an alarm that repeats, you say "Set a repeating alarm for ... at ..." where you include the day of the week you want it to repeat and what time it should repeat ("Set a repeating alarm for Wednesday at 10 am"). When an alarm is going off, you can say "Snooze" to have it wait to go off again for another 9 minutes.

Step Two: Just as with the timers, you can edit it in the Alexa app. You'll select "Timers & Alarms" from the menu, choose the device with your alarm on it, and click on the "Alarms" tab. Here you can select any alarm you want to edit.

Step Three: There is an option next to each alarm called "Repeats." Here you have the following options to choose from: never repeat, every day, weekdays, weekends, every ... (you chose on what day of the week it should repeat), or delete alarm. Once you have chosen an option, you will need to select "Save Changes" to make sure the option you chose remains with that timer.

Once again, a convenient and simple to use skill with Alexa! If you want to change the sound or volume of any of your timers or alarms, go to the tab of the one you want to change (want to change the timer sound? Go to the "Timers" tab) and select "Manage ... volume." With an alarm, you can choose a single alarm in order to change the alarm sound.

I am also including a quick little blurb within this section because this is where it best fits: You can ask Alexa for the current time in your area (if you have the address saved in your Alexa app) and you can even ask for the time in a different area! ("Alexa, what time is it in London, England?") It's a pretty cool feature, and makes it more convenient for you to check the time, especially if you want to call a friend or family member that lives in a different time zone from yourself.

I think that's enough talk about timers and alarms now. Let's move onto how you can make lists.

Managing and Creating Lists

Just as with you timers and alarms, you can have up to a 100 items on a single list. Each item on your list has a character limit of 256 characters, which should be plenty of room for you to make a to-do list or explain each task you need done that day. As a side note, if multiple people in your home use Alexa, there isn't currently a way to create lists for different household members. This means they can edit and add onto your lists if they aren't paying attention. As always, you can access these lists through the Alexa app. You can access your shopping list on Amazon's website as well.

Now that there is background information, here is how you can manage your lists made through Alexa:

- To add an item, say either "Add… to my Shopping List" or "Put … on my To-Do List."

- To check your lists, say "What's on my Shopping List?" or "What's on my To-Do List?"

- To print out a list, use the Alexa app or Amazon website on your computer, select your list, and print it out. On the app, there is an option to print. On the website, you'll have to print the webpage.

- To open a list in the Alexa app, select your left navigation menu and then "Shopping & To-do Lists." To open an app on the Amazon website or Amazon app, select "Your Lists" and then "Alexa Shopping List."

- To add an item, you will select the list you want to add to, type up what you want to add, and select the "Add" or "+" button.

- To edit an item, select an item already on the list, type to edit it, and then select "Save."

- To remove an item, select the checkbox or downward facing arrow next to that item and choose "Delete." You can delete multiple items at once by selecting the checkboxes next to each item you want to delete and selecting "Delete selected."

- To mark an item as complete, select the checkbox next to an item.

And there you go! This isn't as accessible purely through your voice as other skills are, but it can be handy if you need a review of what you need to do each day.

Placing an Order with Alexa

This is definitely something you need to be careful about, and I'll explain why shortly. But you can ask Alexa to place an order for music or for physical products that are eligible to be bought through Prime.

When you make a purchase request, Alexa will search through your order history for Prime-eligible items, Amazon's Choice products – which are highly-rated and well-priced products – and it will also search among other Prime-eligible items.

If the item you are trying to purchase is available, Alexa will tell you the name and price of that item, as well as the estimated delivery time if it will take longer than the typical two-day Prime shipping. If you want additional details about it, check within the Alexa app. Alexa will then ask you to confirm or cancel the order.

If Alexa can't find the item or can't finish the order for some reason, Alexa will offer to add the item to your cart on Amazon, add it to your Alexa Shopping List, or to check the Alexa app for more options.

In addition, you can cancel an order immediately after you place it or track your order once it has shipped through Alexa as well.

Whenever you place an order, Alexa will use the default one-click payment settings on your Amazon account. Thankfully, any orders placed for physical items are eligible for free returns. I am saying this is a good thing in case any children in your home order an item off the Internet on their own. If you haven't heard of it, a child in Dallas ordered 4 pounds' worth of cookies and a $162 dollhouse through Alexa. It shows how easy it is to use, but it may be a good idea to check into the app every now and again to make sure you don't run into any mishaps like this!

There are also requirements that must be met to order anything through Alexa. To order music or a physical item, you need: an Amazon Prime membership (you can make use of the 30-day free trial here also), a payment method that's issued from a US bank and has a US billing address, an Amazon account, and a device that can access Alexa Voice Service. If you want to add a product to your shopping cart or track a recently shipped order, all you need is an Amazon account.

You can, of course, use your Amazon Echo or Amazon Echo Dot in other countries. But these are the basic guidelines for purchasing that I have found based on the Amazon website, so it may require a little more research for the specific country that you live in.

Any music you have purchased through the Digital Music Store will be stored in your Amazon Music library for free! It won't count against any of your Amazon Drive storage limits and will always be available to play or download on other compatible devices that are registered to your Amazon account.

You can also access and update your voice purchasing settings through the Alexa app. Voice purchasing is automatically on your Echo devices, so if you want to turn it off or change it, here's how you can do it:

Step One: In the Alexa app, open your left navigation panel and select "Settings." Then select the "Voice Purchasing" option.

Step Two: Choose the setting you wish to update. You have three options: Purchase by voice (toggling this option will turn voice purchasing off or on), Require confirmation code (where you create a four-digit code and select "Save Changes"—this means each time you voice purchase an item, you will need to say your code, which will not be recorded—and finally, Manage 1-Click Settings (here you simply update your payment method and billing address).

If you have a child in your home, the first two options are most likely to be worth looking into to avoid any incidents like the child ordering herself a large dollhouse and several pounds of cookies.

And there you have it! That's all the basic information I will be providing about how you can buy an item through Alexa on your Echo or Echo Dot. I have covered a lot of basic skills so far, so now let's go into how you can use smart home devices with Alexa.

Smart Home Devices

First off, I know it may seem to be a bit strange to put this information at the end of the chapter. But I thought that this would be the best placement because it does cover a lot of information and many consumers may not have smart home items. Everything else has involved something that almost everyone who has an Amazon Echo device would have. In any case, let me provide information about smart home devices you can use with Alexa!

Of course, not every device can connect with Alexa. There are so many smart devices available (start kits, lighting, thermostats, etc.) that it would take far too long for me to write up each device that is compatible with Alexa. What you should know about your Alexa and Dot is that you can constantly add devices to make your home a smart home. Imagine that it is getting cold in your home at night. You are in the living room and you want your bedroom heat to be turned up. You don't even have to walk upstairs. You can ask Alexa to turn up the heat if you have the right thermostat. Going on vacation? Don't worry about all of your lights being off while you are gone. You can set a timer with Alexa to turn your lights on and off at specific times. And it isn't hard to find the right thermostat, light kit, even whole house kits. Every day, Alexa becomes more popular and this leads manufacturers to create products that will increase the intelligence of your home at an affordable cost.

There are some safety guidelines you should follow when you use a smart home device with Alexa. This is because Alexa will respond to anyone's voice, not just your own, so that could become a safety hazard if you have Alexa hooked up to a door lock or something of that nature. There are two basic things you can do to make sure everything is going well with your smart devise: One, after you make a request, make sure the action was actually completed (for example, if you are locking the door with Alexa, just make sure it did actually lock); and two, make sure you turn off the microphones, and you should probably turn the device off entirely as well, whenever you are away from home, no matter how long you will be gone.

Now that some of the basics have been covered, here are the steps you will need to take to connect a smart home device to Alexa:

Step One: Make sure that you have the manufacturer's app for your smart home device as well as your Alexa app before doing anything else. Once that is done, go to the menu in your Alexa app and select "Smart Home." Then select "Get More Smart Home Skills."

Step Two: Browse for a skill that is appropriate for your smart home service and click on "Enable skill" once you find it. There is a chance that your smart home device can be discovered without a skill, although they will need to be controlled by a skill. To discover them, simply say "Discover devices."

Step Three: You may be prompted to sign in to account that is associated with your smart home device. You will need to follow the prompts afterwards to complete the setup.

If you decide later on to remove a smart home device from Alexa, go back to the "Smart Home" menu, select "Devices," and then click on "Forget" for each device that you want to disconnect from Alexa.

Now that you know how to add and remove a smart device, let's move on to how to create a smart home device group. This is handy if you want to control all the devices within that group at once. It's recommended to only keep like objects together (don't mix your door locks into the same group as your house lights).

To create a group, you go to the "Smart Home" menu, select "Groups," and click on "Create group." Here you can enter the name of the group. Make it easy for Alexa to figure out what you are saying, such as "Bedroom" or "Kitchen." Next, you select the smart home device you want to put into the group and select "Add." Easy as that!

To edit a smart home group, select the group you want to edit from the "Groups" menu. You can change the name by selecting the text field and updating it. You can add or remove a device from this group by selecting the appropriate checkboxes next to that device. You can also delete an entire group by simply selecting "Delete."

So here we are now, at the important part of this section: The commands you use for your smart home device through Alexa! Please note that some of the smart home skills will require you to say "Open..." (you'll have to know the name of the skill for your smart home device in order to open it here) before you can actually make a request. Double-check the skill's detail page to see if this is necessary.

- To turn your smart home device off or on, say "Turn (on or off) ..." You can either call it by the device name or by the group name.

- To activate or deactivate a scene (which is a configuration previously used for the device), say "Turn (on or off) ..." You can either use this by group name or scene name ("Turn on bedtime").

- To set the brightness of your connected lights, say either "Set... to ...%" ("Set bedroom to 60 %") or "(Brighten or dim) ..." ("Brighten living room").

- To control a thermostat with Alexa, use the command "Set... temperature to ... degrees" ("Set hallway temperature to 72 degrees") or "(Increase or decrease) the ... temperature" ("Decrease the Nest temperature").

- To check on your thermostat, simply ask "What's the temperature in here?" or "What's my thermostat set to?"

- To change the fan speed, say something along the lines of "Set my fan to ...%" ("Set my living room fan to 40 %").

Finding Local Restaurants and Stores near You

This is handy if you are looking for a new place to go within your area. It may not include information on every nearby location, but this will provide a lot of very easy-to-access information. Before you do anything else, you will have to add your address to the Alexa app.

Once that has been done, you can ask questions like "Alexa, what bakeries are nearby?" or "Alexa, find the hours of a nearby grocery store." Alexa will then look up and provide all the information pertaining to these questions to you. It's pretty simple to use, and it's a little easier than staring at a map in Google Maps and randomly selecting one.

Depending on your location, you can even order dinner delivered to you! Many nationwide companies are offering food delivery options through the Alexa app. If, say, pizza delivery is available in your area, you can say "Alexa, order a pizza through (the company's name)" and you will have pizza delivered to your home.

Check the Weather

This is very simple to do, as long as you have your address in the Alexa app. All you have to do is ask is "What's the weather?" to check for the weather at that moment. If you want to know about a future forecast, you can ask something like "What's the weather for this week?" or "Will it rain tomorrow?"

Additionally, you can ask about the weather in a different area from you, the same way you can ask about the time in a different area. Just ask "What's the weather in..." ("What's the weather in New York City?") to get your information.

Third-Party Applications

This isn't something I have particularly made note of so far, but the reason as to why and how there are thousands of Alexa skills available for you to use is due to allowing outside parties to create an app especially for Alexa-run devices. This isn't a particularly important section if you just want to know how to run your device, but I thought it would be good to add in case anyone wants to make an app.

Amazon has also made it a bit easier for everyone to create an app – and also have actually *made* an incentive for people to create an app (you can get a free Alexa developer t-shirt or sweatshirt!) Creating an app can be done by following six steps. I won't go into them here, but it is relatively easy to make a skill the way Amazon has set it up. You can find the instructions to do so on Amazon's support page.

This is also relevant because Alexa supports something called IFTTT, or "If This, Then That." This is a third-party service that allows you to automate how your devices work with one another! You do have to be signed up for this service and you need to enable through your Alexa skills. But this is a good skill to have to do things like...

Finding Your Phone

At one point or another, it's likely that you have misplaced your phone around the house and needed it so you could get out the door and arrive at some appointment on time. If you have an Echo device, you can easily locate your phone! You do this by installing the TrackR app on your phone and enabling the Alexa skill. You will, of course, have to have done this while you know where your phone is.

To find your phone, just ask Alexa, "Alexa, where's my phone?" This will make your phone ring very loudly so that you can locate it, even if it is in silent mode. It's very convenient if you are in a rush to get out the door right then and there.

Additional Information

So, as mentioned at the beginning of this chapter, I am not going to fully explain everything on the list of what you can do. But this can become very handy for when you are busy working on something. As I mentioned within the section with audiobooks in this chapter, you can have Alexa read recipes from a cookbook for you!

You can also ask Alexa things like "How many cups are in a quart?" while you are cooking and set times so that you won't have to look it up for yourself as you are trying to cook. The same thing applies if you are asking Alexa about a math equation. Never know when you'll need that if you are a student or if you work in a very math-heavy field. You can use this same kind of feature whenever you need to settle an argument. Can't agree on how to spell a word or who the lead actress was in a particular movie? Ask Alexa.

And if you miss something she said or need her to repeat that answer, just ask "Alexa, can you repeat that?" She'll repeat the same information to you as many times as you need it. You can also access the information from the Home screen of the Alexa app by looking through the cards.

If you add your location to Alexa, you can even ask about what traffic looks on the way there. Simply ask "Alexa, how is traffic?" or "Alexa, what's my commute?"

Want to entertain yourself and your friends while they are over? Play a game! You'll likely have to enable the skill, but it sounds like fun to have a voice play-through of "The Wayne Investigation" (a murder mystery game) or to play something like "Movie Quotes" (matching the line to the right movie). This can be done by saying, "Alexa, start ..." ("Alexa, start The Wayne Investigation."

Another feature, similar to the "Things to Try" menu within the app, is that you can ask Alexa this question: "Alexa, what new features do you have?" Alexa will explain whatever new skill was added recently. This can be an interesting way to find new skills you want to use while you are making use of your Echo devices.

The last feature I am going to make note of is that Alexa not only can answer your questions, but also tell you a joke or show references! Use a popular quote from your favorite franchise (like "Winter is coming" or "Alexa, beam me up") and she'll respond with a quote or some information about the series. Other Easter eggs in Alexa include how she responds to someone asking if Santa is real or how babies are made (you never know when your children will ask, after all). Another interesting Easter egg (one of many, I'm sure) is that you can ask Alex to flip a coin! Ever need to decide on something with a coin flip right then? Just ask your Echo device! So try it out and see what Alexa will say!

And there you have it! That's a lot of the more basic information I can provide on Alexa that you, as a user of an Amazon Echo or Amazon Echo Dot, are mostly likely to make use of regularly. Amazon is littered with guides on how to use Alexa and all of the skills and commands that come with them. So if you want to use a skill that I haven't listed, there is a ton of information out there, provided by the experts at Amazon themselves!

But, as with every technological item in existence, you can still run into problems and issue with Alexa. Here is a basic rundown on what common issues you may run into and how you can fix them.

Problems Streaming

If you find that you can't stream any music or audiobooks from Alexa, here are a few things to check and try:

- Keep in mind that a poor Internet connection and low available bandwidth are the usual reasons for streaming issues. If this is the problem, turn off devices you aren't currently using and move your Echo device closer to the router and away from sources of interference, such as microwaves.

- Many Wi-Fi devices will automatically connect to the 2GHz channel on your router. If you have a dual-band router, you can connect your Echo to the 5GHz channel instead. Here, there will be much less interference and you'll have better range.

- Try restarting your Echo device, router, or Internet modem. This may solve a lot of the problems you are having. Amazon has suggested that you turn off your router and modem and then wait 30 seconds before turning the modem back on. Once the modem has restarted, turn the router back on. Once the router has restarted, turn your Echo off and on again.

- If you have a firewall on your network, check to see if the following UDP ports are open: 123, 443, 4070, 5353, 40317, 49317, and 33434. If you don't know if you have a firewall, contact the network administrator to check. You should also be able to ask them this question.

- If none of these options work, then you should contact your Internet service provider, your router manufacturer, or your network administrator to see if one of them can help you to fix the issue.

Bluetooth Connectivity Issues

There are a few different reasons why your Echo or Echo Dot may not connect with your Bluetooth-enabled device. Here are a couple of ways that you can try to fix the problem:

- Check the batteries or charge of your Bluetooth device. They won't be able to connect if they don't have enough power.

- Move your Bluetooth device and your Echo device away from wireless devices that may cause interference. You should also make sure it is near your Alexa device when you pair the two together.

There's one more thing you can try, but it is a bit of an intensive process, so I am separating it from the rest of the list. This involves clearing all Bluetooth devices from your Echo device.

For your Echo, this can be done by going to the "Settings" menu in your Alexa app, selecting your Echo, and then clicking on "Bluetooth." This will give you the option "Clear" to remove all Bluetooth devices from the Echo.

For your Echo Dot, go to the "Settings" menu, select your Echo Dot, and click on "Bluetooth." You have to select a Bluetooth device from the list and then click on "Forget." To remove all the devices, you must repeat this for every device, as there is no mass "Forget" option available. Once this is done, restart your devices.

To pair your Bluetooth device with your Echo again, open the setting menu on your mobile device and make sure Bluetooth is turned on. You should be near your Echo as you do this. Once it is on, say "Pair." This will make your Echo enter pairing mode. On your mobile device, go into the Bluetooth setting menu and select your Echo. Alexa should then tell you if the two devices successfully connected.

To pair your Bluetooth device with your Echo Dot, open the Alexa app and go to the "Settings" menu. Select your Echo Dot, then press "Bluetooth." This should lead to an option called "Pair a New Device. This will make your Echo Dot go into pairing mode. On your mobile device, go into the Bluetooth setting menu and select your Echo Dot. Alexa should then tell you if the two devices successfully connected.

Those are the best solutions available if you are unable to connect a Bluetooth device with your Alexa-controlled device. If none of these work, you can try contacting customer support for further assistance.

Alexa Can't Discover My Smart Home Device

The first thing I suggest checking is whether your smart home device is actually compatible with your Echo devices. If no, it would be incredibly difficult to connect Alexa with your smart home device. The second thing you should do is see if a skill is required for your device or not. If neither of these was the problem, check out the tips below:

- Make sure you have downloaded the app that goes with your smart home device and then set it up.

- Restart your Echo device and your smart home device.

- Try disabling and re-enabling the smart home skill within the Alexa app.

- Make sure you have all of the latest software updates installed for all of your devices.

- Make sure your Echo device and your smart home devices are on the same Wi-Fi network. If you are using a public Wi-Fi connection, such as one at your workplace, then they may not be able to connect to one another.

- You may need to contact your router manufacturer for help here, but you can also go on your computer to turn on SSDP/UPnP on your router.

- Check the group name you've assigned to the smart home devices. The name you assign to it needs to be easily recognizable to Alexa, so check the spelling of the group name.

- If you need to discover you device again, say "Discover Devices." Once the discovery process has been completed, Alexa will tell you whether or not any devices were found.

There is a chance that your smart home device doesn't need a skill to be discovered by Alexa. If it is a Phillips or Wemo device, it will not need a skill. To discover Philips devices, hit the button is on your smart device's bridge. Then tell Alexa to "Discover devices." For Wemo devices, all you need to do is say "Discover devices."

That's all I have for information on how each device works and functions, as well as a few fixes for problems you may run into. Now it is time to go into different reviews on what others have said about these two Amazon products, as well as questions that are often stumbled upon along the way.

Peter Gelson

Chapter 7
Amazon Echo FAQs and Official Reviews

I know this may be a bit confusing if you have looked at the table of contents. I have done everything basically the same way up to this point, so why am I changing it now? This is because many of the questions asked about the Echo will also be applicable to the Echo Dot. I see no need to repeat the same exact information in two separate chapters. So this chapter will include FAQs about the Echo devices as well as reviews done by various news outlets on the Amazon Echo. The next chapter will have a summary of the reviews done by the same news outlets on the Amazon Echo Dot.

Now that that's been cleared up, let's begin!

FAQs on Echo Devices

1. How do the Echo devices recognize the wake word? How do I know when my voice begins being streamed after it wakes up?

There is a system of keyword spotting that they do in order to detect the wake word. Once the wake word has been detected, they stream audio from there to the cloud, in addition to a fraction of a second's worth of audio before you say the wake word. The light ring on top of your device will also turn blue to notify you that your voice is, in fact, being streamed right then.

2. Can I turn off the microphones so all my conversations won't be recorded?

Yes, you can! There's a button right on top for that very thing. It should be noted that, while your microphone is always on, everything you say isn't being recorded to the cloud. It only records when you make a request so that it can improve itself and work more efficiently.

As I mentioned before, there weren't very many questions I could find that were widespread. So let me instead go into the next section of this chapter.

Reviews

The review I am going to begin with is by Farhad Manjoo. He wrote an article that was published on the *New York Times* website called "The Echo from Amazon Brims with Groundbreaking Promise." He begins his review by saying that technological companies have been working hard to improve and create devices to become the new device that is as widely wanted as a smartphone. He goes on to say that the Echo may be the new widely desired device, based on how it has snuck into his life.

He also notes that Amazon has managed to bypass Apple and Google by going straight for something that remains in the home, where a lot of time is spent. It is already being bought by enough consumers that it is consistently going out of stock every few weeks, at least according to Scot Wingo, who is the chairman of an e-commerce consulting firm called ChannelAdvisor. Manjoo finishes his review by saying that the Amazon Echo is a lot better suited to home usage because it means you do not even have to type something into a screen. He believes it can develop greatly and overall seems to really enjoy the device.

Next up is a review from *The Guardian*'s website. It's titled "Amazon Echo review: the best combined speaker and voice assistant in the UK" by Samuel Gibbs. It took longer for the Echo to come out in the UK than it did in the US, so Gibbs started his reviewing wondering if it was worth the wait. Just like Manjoo, the longer he had the Echo in his home, the more accustomed he became to having it around. He noted one of the most useful perks in his life was when he needed a calculation for a project he was working on. All he had to do was shout across the room instead of dropping all the tools in his hand and he could continue working.

He finished his review with a list of the Echo's pros and cons. The pros included the fact that the Echo can almost always hear you no matter where you are standing in comparison with it, it has a mute button to disconnect the mics, the light ring is incredibly obvious and useful, it's full of third-party apps so you don't have to only use Amazon apps, the speaker itself sounds pretty good, and, one of the things he really enjoyed, was how Alexa sounds far more human than other voice assistant systems like Siri. The cons included that the Echo can't always answer the questions you ask it, you will have a device that is constantly listening for your voice within your home, Alexa doesn't support multiple user calendars or personal information yet, and that only one Spotify or Amazon music account can be linked at any given moment.

The next review is from pcmag.com, where Alex Colon and Will Greenwald reviewed the Amazon Echo. They note that the Echo is very good at being a stationary speaker – so long as you don't turn the volume too high or too low, which is when sound distortion begins – and a disadvantage is that it can't currently handle information like reading your emails or text messages to you. They finish off their review by saying that the Amazon Echo isn't a good choice if you just want a speaker, but it functions wonderfully as a voice assistant for you.

Then there is a review from cnet.com, which was written by Ry Crist and David Carnoy. They gave the Amazon Echo an overall 8.3 out of 10. The lowest scoring section, they believed, was performance. They went into detail to describe how the Echo works and what it can do. They also mentioned the things that the Echo – in its current form – cannot do. For example, it can't connect with external speakers or smart TVs. It doesn't allow you to call someone through the Echo or to take calls. But, as it is in its current form, it is very helpful and can connect and do many things for a single device. The review finishes by mentioning how much the Echo could assist the elderly and disabled by allowing them, especially those who have minimal amounts of movement, if any, to have more independence in controlling their environment.

Finally, engadget.com wrote a review on the product and mentioned that the Amazon Echo is a good product for people who are already signed up with several Amazon services. They note that this is definitely not the product for someone just looking for a speaker to use. They also note that you can't easily carry around the Amazon Echo around with you. Not because of the size of the device, but because the Echo needs to be plugged in to work. For their rating system, they combined several reviews and ratings together to create an "Average Critic Score." This score was a 7.7 out of 10. Following this rating score were audio quality (rated 7.6), design and form factor (rated an 8), and portability (rated a low 5.3).

There are over 50,000 reviews of the device on Amazon, too many to go through and read for a well-written but concise review to include within this book. So instead of including some in here, I will let you know that the Amazon Echo currently has a 4.4 out of 5-star rating, with 67% of users giving it five stars.

Hopefully these reviews help put what the Echo does into perspective – at least slightly – and let you know how happy you will be with your device. (You know, in case you're having second thoughts about your US$180 product and may want to return it instead.)

In any case, the reviews on the Amazon Echo are now finished. It's time to move on to the reviews about the Amazon Echo Dot (both the first-generation and second-generation!). (Amazon Echo Review The Best Combined Speaker and Voice Assistant in the UK.)

Peter Gelson

Chapter 8
Amazon Echo Dot Official Reviews

As mentioned in the last chapter, the FAQs for both Echo devices are exactly the same. However, the reviews will say different things about each product, as they work differently in some ways. So let's begin!

First up is another review from cnet.com on the first-generation Echo Dot. I'll also include their review on the second-generation Echo Dot as there are some differences. To start off the review, ratings were given for its features, usability, design, and performance. It got a 9 out of 10 for features, 8 for usability, 10 for design, and 8 for performance. Averaged out, it was given an 8.8 out of 10.

To start off the review itself, Crist explains that the Dot works the same way the Echo does, but it is just quieter. This is balanced out by the ability to connect it to another set of speakers. It's a very small edition of the Echo that is well worth it, considering it is half the price of the Echo. A downside is that it takes much longer for the Echo Dot to respond to its wake word when it is using the speakers built into it than the Echo does. He adds that the Echo Dot is still advertised as a Prime-only device that you can only buy through Alexa. You can however, still order the Echo Dot by using the microphone within the Amazon app on your phone and saying "Add Echo Dot to shopping cart."

Ry Crist also reviewed the second-generation Echo Dot. As with the first-generation version, it was first rated upon features, usability, design, and performance. The second-generation Echo Dot scored higher in almost every feature; it got a 10 for features, 9 for usability, 9 for design, and 10 for performance. Overall, it got a 9.5 out of 10 rating! Just as with the first Echo Dot, they have made note that you can connect the Dot up to external speakers. Unlike the first-generation device, however, it doesn't come with a cable for you to be able to set it up with your external speakers. You'll have to get one separately. One major difference between the two generations was that the second-generation Dot could hear much better over the sound of music playing – whether it was close to a set of external speakers or if it was blasting music from its own internal speakers. Not only did this make the new Echo Dot much better for Crist, but also the fact that you can buy this version of the Echo Dot for $50.

From *The Guardian*'s website, Samuel Gibbs reviews the device as being the part that makes an Echo interesting without the speaker and costing much less, as it is priced at US$50. They went into detail about the specifics of Alexa and the Echo Dot, as well as how they can be used. Gibbs mentions that the Echo Dot can blend in well with your home because of the size it is. You can get the device in either all black or white with a black top, which may make it stand out more. Plus you can get cases for the Echo Dots. Some of the pros he includes in his review are the fact that it can almost always hear you, it can easily be heard despite its size, and that you can use voice control for many different functions. Some cons he included were that it really should be connected to an external speaker for music, it can't always answer your questions, and it doesn't support multiple Spotify or Amazon music accounts at once.

Roberto Baldwin from engadget.com reviewed the Amazon Echo Dot and liked it much better than the Echo (as can be seen from the title of the review, "Forget the Echo. Buy [the Echo Dot] instead." As it is a new feature for an Echo device, Baldwin spends time talking about the speakers are on the Echo Dot. He says that they work and sound fine, but having a stereo system or using one you already own is way better.

He makes note of how inconvenient and odd it is to not be able to use the Echo devices together to play the same music or share the same tasks. If you use the same wake word for both, there is still a possibility that both will answer, as Baldwin found out during his testing of the Echo Dot. Baldwin concludes his review by saying that, while the Echo Dot and Alexa app aren't without faults, it's a very good value for only $50 and can help make finding out and checking information much easier.

Michael Brown from techhive.com reviewed the Echo Dot and said that this would be the best one for most people to buy out of all of the Echo devices. He makes note of the fact that the Echo's mics aren't very good at filtering out outside sounds, like noise from the TV or having music playing loudly. He said that he would just walk closer to the Echo or pause the TV. He does note that the Google Home, a device similar to the Echo, is much better at filtering out that outside noise and can listen in much more easily. Keeping that in mind, he goes on to say that the newest Echo Dot had better mics than the one in the past, although it still had issues with outside noise at times.

He himself has a smart home – with 32 smart dimmers and switches and three smart entry locks, among many others – and connects and uses many things with his Echo Dots. He actually placed an Echo Dot in his garage, partially to counteract the fact that there wasn't a light switch installed next to the door that leads to the front porch. Instead of stumbling through the dark, he can control the lights and turn them on as he walks in. He also notes that Alexa can't unlock any of your smart locks or things like that, just close them. Overall, he really enjoyed the device and said it is very cheap for a hands-free device that does as much as the Dot does. He does mention that Alexa is less sophisticated than Google Assistant, as Alexa has a weaker speech recognition system currently in place.

Those are all of the reviews I am going to include on the two generations of Echo Dots. I think these reviews provide a good explanation as to how well an Echo dot holds up in comparison with other devices and how well they truly work! Now it is time to move into the final chapter of this book.

Peter Gelson

Chapter 9
Alexa FAQs

There were a few questions I found that I hadn't answered elsewhere in this book. I hope this will help round out everything you have learned!

1. Will Alexa's voice recognition improve over time?

I did touch on this in the book, but I'll reiterate it here. Alexa will use your voice recordings and other information it gains through the Cloud in order to improve your experience using it. Alexa will process information such as your music playlists in order to fulfil your requests. When you use the voice training option in the Alexa app or provide feedback on how well interpreted your response was, it will also help to improve the voice service!

2. What are the return policies for products purchased through Alexa?

There are many policies for each different area of products that Amazon sells, so that would be a bit long to get into.

Peter Gelson

Check online to see what the rules are for returning an item as it is different for almost every type of product (handmade, Amazon devices, collectables, etc.).

If you accidently purchase a song or album from the Digital Music Store with Alexa, however, you can get a refund, as long as you request a refund within seven days from the time you bought it. Make sure to contact customer service for anything dealing with your refunds.

3. What are some of the best/funniest tricks you can do with Alexa?

I also went into this briefly, but there are so many things you can discover and do with Alexa that I am going to include another small list here of some of the funniest or best things you can use Alexa for! I would also like to note that these are all especially funny, as you have to imagine the Amazon programmers thinking of good references and jokes to include in the device.

- "Alexa, how many licks does it take to get to the center of a tootsie pop?"

- "Alexa, do aliens exist?"

- "Alexa, do you want to build a snowman?"

- "Alexa, do you have a boyfriend?"

- "Alexa, how tall are you?"

- "Alexa, what does the fox say?" (There are several answers you can get by asking this – have fun!)

- "Alexa, rock paper scissors."

- "Alexa, my name is Inigo Montoya."

- "Alexa, tell me a joke."

- "Alexa, define supercalifragilisticexpialodocious."

- "Alexa, Up Up, Down Down, Left Right, Left Right, B, A, Start" (this is the longest joke users have found – so far)

- "Alexa, do you dream?"

- "Alexa, do a barrel roll."

- "Alexa, who's better, you or Siri?"

- "Alexa, sing ..." (you can ask her to sing you a song, just include the title of the sing in your query)

There are so many more you can find out there, all with great and hilarious answers. If you want even a sample of the potential answers, here are a few I'll share. Find the rest out on your own to pass the time or break in your new Echo device!

"Alexa, are you friends with Glados?" "We don't really talk after what happened."

"Alexa, who let the dogs out?" "Dog walking isn't my area" OR "I don't know, but they seemed happy. "

"Alexa, how many roads must a man walk down?" "The answer, my friend, is blowing in the wind."

"Alexa, what is the meaning of life?" "The answer is 42, but the question is more complicated."

One more thing I'll include here: Want to pull a prank with your Echo device and you have a voice remote? Go into another room (where your friend or family won't be able to hear you) and use the remote to make Alexa say what you want it to say using the "Simon Says" feature. Want to intensify the game? Refer to each person in the room as they talk by their voice so it seems as if Alexa is responding. It may not work for long, but it could still be a funny prank to pull.

Peter Gelson

Conclusion

Thank you very much for taking the time to read this guide book for the Amazon Echo devices! Take this knowledge with you as you continue to learn the ins and outs of your Amazon Echo devices. There's quite a lot of them with the thousands of Alexa skills at your disposal, after all!

Now that the book has officially ended, I can only hope that you have gained a lot of knowledge about how to use your Amazon Echo devices! You know how they function, what their physical characteristics are, and many of the things that can be done with them. There was also some information on smart home devices in this book, which is another way technology has been improving in recent years.

I hope that you now see Amazon's latest products as a step to further developing technology. I mentioned in the introduction that technology has already vastly improved from where it was not too long ago. This development of the Amazon Echo devices – as well as other similar devices – can lead to the furthering of technology in our day-to-day lives.

Alexa will also be an interesting way to improve AI and voice recognition technology. Although Alexa isn't *exactly* an AI, the voice recognition technology and the continuous development and adaption to its users could improve the way future technology in this field is used.

In any case, I hope you discovered a lot more about your Amazon Echo or Amazon Echo Dot than you knew before! Enjoy learning about all the things you can use these devices for and using them to help assist you throughout your day. And have fun discovering all the jokes and other references Alexa will report back to you!